HUNTERS

Picture Credits

Stephen Frink: pages 7-8
Glen & Rebecca Grambo: Back cover; pages 10, 26-27
Breck P. Kent: pages 7, 13, 16-17, 19, 29
Dwight Kuhn: Endpages; pages 10-11, 20-21, 21, 24-25, 25, 28-29
Tom & Pat Leeson/DRK: Cover
Zig Leszczynski: Endpages; pages 8-9, 9, 14-15, 15, 17, 18, 28
McNeil River Sanctuary: page 6-7
A.B. Sheldon: pages 24-25
Lynn M. Stone: pages 7, 10-11
Merlin D. Tuttle: page 11
Norbert Wu: pages 16, 24, 29
Michael Fogden/DRK Photo: page 21
Jeff Foott/DRK Photo: page 23
R.F. Ashley/Visuals Unlimited: page 8
Francis & Donna Caldwell/Visuals Unlimited: pages 12-13
Beth Davidow/Visuals Unlimited: page 27
T. Gula/Visuals Unlimited: page 6
Ken Lucas/Visuals Unlimited: page 15
Joe McDonald/Visuals Unlimited: page 14
Kjell B. Sandved/Visuals Unlimited: pages 12-13
Leonard Lee Rue III/Visuals Unlimited: page 27
Will Troyer/Visuals Unlimited: page 9
Bob Bennett/The Wildlife Collection: page 23
Mauricio Handler/The Wildlife Collection: pages 22-23
Martin Harvey/The Wildlife Collection: pages 13, 22-23
Kenneth J. Howard/The Wildlife Collection: pages 18-19
Tim Laman/The Wildlife Collection: page 26

Published by Rourke Publishing LLC
Copyright © 2002 Kidsbooks, Inc.

All rights reserved.
No part of this book may be reproduced or utilized in any form or by any means, electronic or mechanical including photocopying, recording, or by any information storage and retrieval system without permission in writing from the publisher.

Printed in the USA

Grambo, Rebecca
　Hunters / Rebecca L. Grambo
　　p. cm — (Amazing Animals)
　ISBN 1-58952-147-1

HUNTERS

Written By
Rebecca L. Grambo

New Hanover County Public Library
201 Chestnut Street
Wilmington, North Carolina 28401

Rourke Publishing LLC
Vero Beach, Florida 32964
rourkepublishing.com

ON THE PROWL

When we're hungry we look in the kitchen for food, eat in a restaurant, or go to the grocery store. For hungry animals, things are more complicated. Animals that hunt have to catch their food.

This green mantis blends in with the plant on which it waits for prey. As an unsuspecting insect comes close, the mantis snatches it up with strong front legs. Blending with the background is called camouflage (CAM-o-flahj). Many animals use camouflage to fool prey into approaching.

This bear is dining on fresh salmon. Bears like to eat meat and fish as well as plant foods like berries. An animal that eats many kinds of food is called an omnivore (OM-ni-vor). People are definitely omnivores!

Sharks are awesome hunters. They can find their way to prey in the open spaces of the ocean. Sharks rely on their sense of smell. And they are very good at detecting sounds and vibrations. Once sharks are close enough, their eyesight guides the final attack.

The jaguar is a carnivore (CAR-ni-vor). Carnivores mainly eat meat. They work very hard to catch their food. They may fail many times before they are finally successful.

Herons hunt by standing very still. They watch for a tasty fish or frog to swim by. When they see one—ZAP! The strong muscles in the heron's neck push its head forward very quickly. Sometimes the heron strikes so hard that its long bill goes right through the fish!

LAND LOVERS

A variety of animals live and hunt on land. Some tigers hunt in tropical jungles. Others track prey in the snow. But no matter where they live, tigers are hunting for meat. They are carnivores.

Some snakes make other snakes a regular part of their diet. The rattlesnake is poisonous. But that doesn't bother the king snake at all!

The star-nosed mole hunts *under* the land, using its big claws to dig quickly through the dirt. Its weird-looking nose is very sensitive, and helps the mole find its favorite food—earthworms.

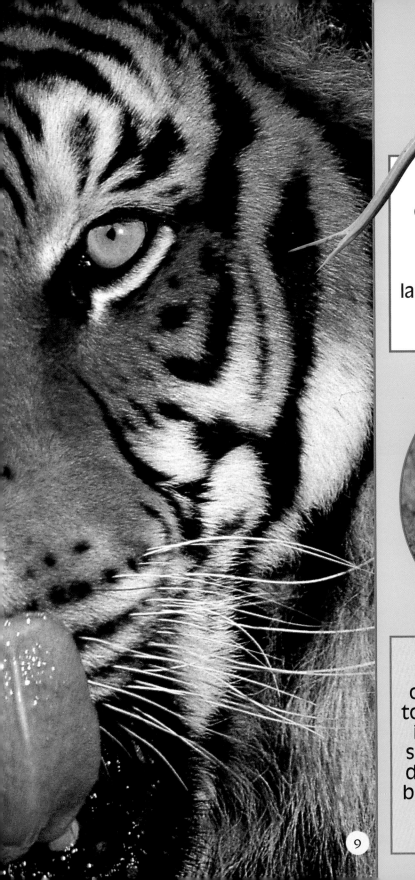

This 300-pound lizard eats wild pigs and deer. Like other reptiles, the Komodo dragon uses its tongue to "smell" the air for food. Often the Komodo dragon lashes prey with its powerful tail before attacking with its jaws.

The secretary bird walks around the hot, dry plains of Africa looking for snakes to eat. It hits the snake with its wings, then stomps the snake until it is dead. If that doesn't work, the secretary bird takes the snake up into the air and drops it to finish it off.

OUT OF THE SKY

Many animals do their hunting in the air. Birds feed themselves and their young by catching insects. The house wren below has captured a fly. Think how quick a bird must be to catch a fly in the air!

The dragonfly is a fierce flying predator. It holds its front legs like a basket to scoop up other insects in midair. The largest dragonflies sometimes snatch up small frogs and fish from the water.

Bald eagles like to eat fish. An eagle may find a fish that is already dead. Or it may dive down and grab one from the water. The bald eagle may even steal its meal from another fishing bird by scaring it into dropping its catch. Then the eagle flies off with lunch!

The fishing bat does its airborne hunting at night. It uses sound to help it find food. The bat sends out a series of very high-pitched clicks. The sounds bounce off objects telling the bat about its surroundings. The bat can even detect ripples made by a fish or frog in the water. It then scoops up its prey and flies away. And all this takes place in the dark.

WATER WORLD

Lakes, rivers and oceans serve as hunting ground for many different predators.

Piranhas (puh-RAH-nas) live in the rivers of South America. They are small but have razor-sharp teeth. Piranhas hunt in groups called schools. They usually eat other fish. But they may attack a wounded animal or person that enters the river.

It may move slowly but the starfish is still a hunter. A starfish eats clams, scallops, and other starfish. It can even pull a hermit crab out of its shell. Once a starfish has latched on to its prey, it does something strange. The starfish wraps its stomach around the prey and begins digesting its meal—in this case, a mussel.

Puffins nest on land but they do their hunting in the sea. They love to eat fresh fish. When they are chasing fish, puffins almost look as though they are flying underwater.

Crocodiles lurk in the water waiting for prey. Often, only their eyes and nose show above the surface. They may eat fish, as this Nile crocodile is doing. Or they may catch larger animals that have come to the water for a drink.

IN THE DARK

Nocturnal (nok-TURN-ull) predators do their hunting at night. Owls are experts at this. They can see very well in low light and have excellent hearing. An owl can spot prey that people would never notice in the dark. And an owl's wings are built for quiet flight.

With its pigeon-toed walk and flattened shape, a badger looks funny. But there's nothing funny about a badger if you're a ground squirrel! Badgers make nighttime raids on ground squirrel colonies.

The fennec fox does its hunting in the deserts of Africa. Its big ears are very sensitive, and can pinpoint the location of prey in the dark.

You may not think of bullfrogs as predators. But at night, near the water's edge, they hunt for food. Some of their favorite meals include fish and other frogs. Large bullfrogs sometimes eat small birds and snakes.

DANGEROUS TEMPTATIONS

If you've ever gone fishing, you probably used bait or a lure to catch fish. Some animals have built-in lures that they use to attract prey.

The anglerfish does its fishing in the deep ocean. On its forehead, it has a thin rod with a bump on the end. The anglerfish sits still and jiggles its lure. Any fish that investigates the lure is quickly gulped down.

This young copperhead is almost hidden in the leaves. The snake is using the yellow tip of its tail as a lure. To the frog, it looks like a worm or something else good to eat. When the frog hops close enough, the snake will strike.

The alligator snapping turtle blends in with the muddy river bottom where it hunts. But on its tongue is a bright piece of pink flesh. The turtle hunts by sitting with its mouth open. A fish, mistaking the pink lure for a worm, swims up for an easy meal. Then the turtle snaps its jaws, and the fish becomes turtle food.

DOOM IN DISGUISE

Hunting is easier with the right kind of disguise. Many snakes that do their hunting up in the trees are green. They blend in with their leafy surroundings. This snake may be hunting for lizards or birds.

For insects, a beautiful flower may hide danger. There are some spiders that do their hunting on flowers. They are camouflaged to look like the flowers where they live. The spiders sit very still and wait for their prey to land.

A transparent sea-jelly is very hard for prey to see and avoid. Seajellies hunt while drifting in the sea. They trail stinging tentacles beneath them. The poison in the tentacles can paralyze small fish. Then the jelly slowly absorbs the small fish into its system.

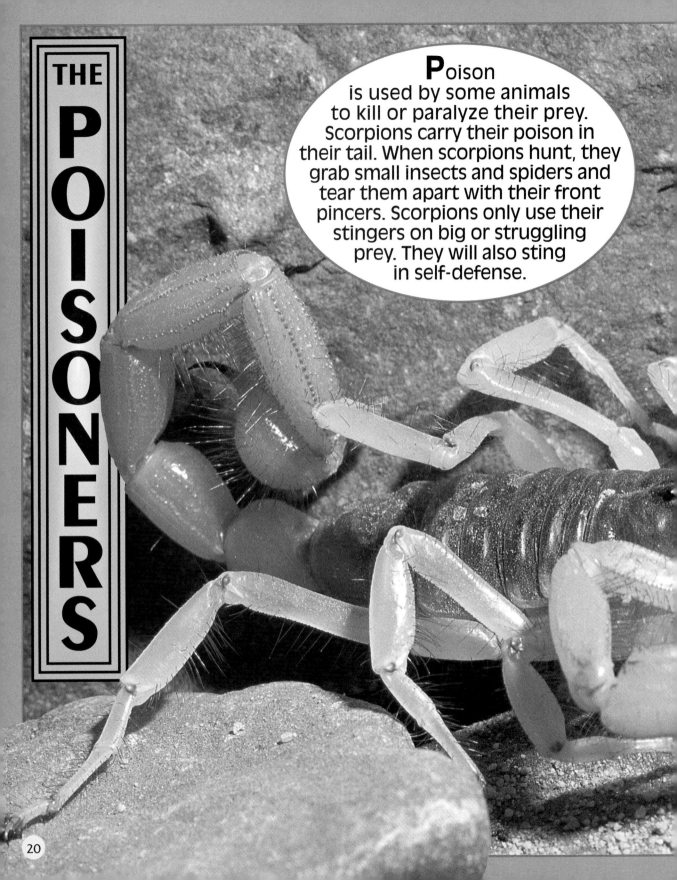

THE POISONERS

Poison is used by some animals to kill or paralyze their prey. Scorpions carry their poison in their tail. When scorpions hunt, they grab small insects and spiders and tear them apart with their front pincers. Scorpions only use their stingers on big or struggling prey. They will also sting in self-defense.

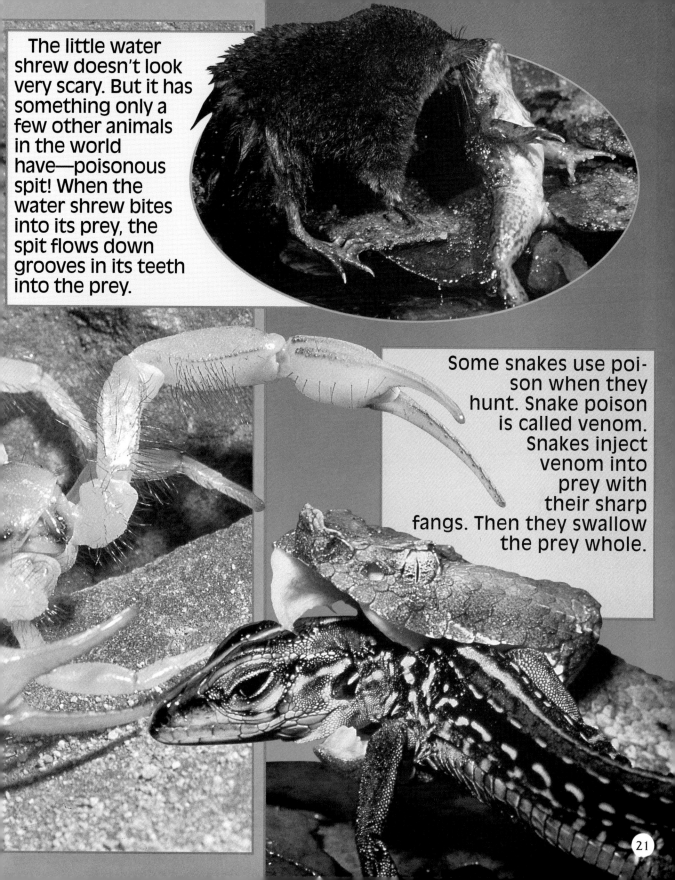

The little water shrew doesn't look very scary. But it has something only a few other animals in the world have—poisonous spit! When the water shrew bites into its prey, the spit flows down grooves in its teeth into the prey.

Some snakes use poison when they hunt. Snake poison is called venom. Snakes inject venom into prey with their sharp fangs. Then they swallow the prey whole.

FAST AND DANGEROUS

Many animals rely on speed to get a meal. Very sharp teeth in a very fast body—that's the barracuda. Its slender, silvery body is hard for its prey to see. The barracuda is one of the sea's speediest hunters.

The cheetah is the fastest predator on land. It can run almost 60 miles an hour! But it can't run this fast for very long. The cheetah must catch its prey quickly or give up and wait for another chance.

Martens attack their prey with quick, darting motions. They are notorious for hunting unusual prey. Martens are probably the only animals that regularly hunt porcupines for food.

Robber flies are quick enough to catch their food on the wing. The robber fly sits and watches for other insects to pass by. When it spots one, the robber fly dashes off and snatches its prey in midair.

SPECIAL TACTICS

A secret weapon is a great way for a hunter to get food.

The electric ray has a stunning secret weapon. It can make electricity. When prey comes near, the ray pounces on it and then— ZAP! The stunned prey becomes the ray's dinner.

The archerfish waits underwater, watching for insects on overhanging leaves or grass above the water. When it sees one, the archerfish takes careful aim and spits! If hit, the insect falls into the water. Then the archerfish eats it.

The chameleon (kuh-MEAL-ee-un) is a lizard that likes to eat insects. A fly may think it is far enough away from the chameleon to be safe. But the chameleon has an unpleasant surprise—an extra-long, sticky tongue. Some chameleons can stretch their tongue out for a distance that is longer than the length of their whole body, including their tail!

Spiders set a trap for passing insects. They build webs with silk made in their body. The silk threads are sticky. When an insect gets caught in the web, the spider wraps it in silk and gives it a poisonous bite to paralyze it.

Not all spiders use their silk just to make webs. This jumping spider has spun a safety line. If the spider misses its prey, the line will stop it from falling to the ground.

PARTY OF HUNTERS

You may already know that lions work as a pride, and wolves work as a pack. Some other animals also hunt in groups.

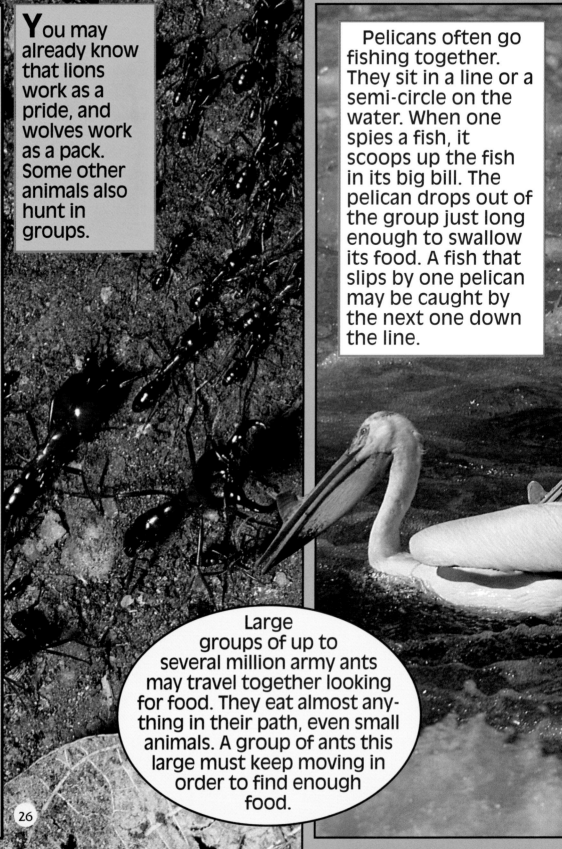

Pelicans often go fishing together. They sit in a line or a semi-circle on the water. When one spies a fish, it scoops up the fish in its big bill. The pelican drops out of the group just long enough to swallow its food. A fish that slips by one pelican may be caught by the next one down the line.

Large groups of up to several million army ants may travel together looking for food. They eat almost anything in their path, even small animals. A group of ants this large must keep moving in order to find enough food.

Wild dogs hunt in packs. Working together, they can bring down much bigger game than a single dog could. And there is less chance that the prey will escape.

Orcas are sometimes called killer whales. They live and hunt in groups called pods. Orcas eat seals and sea lions, penguins, squid, and fish. Working as a group, orcas will also attack other, larger whales.

SUCKERS!

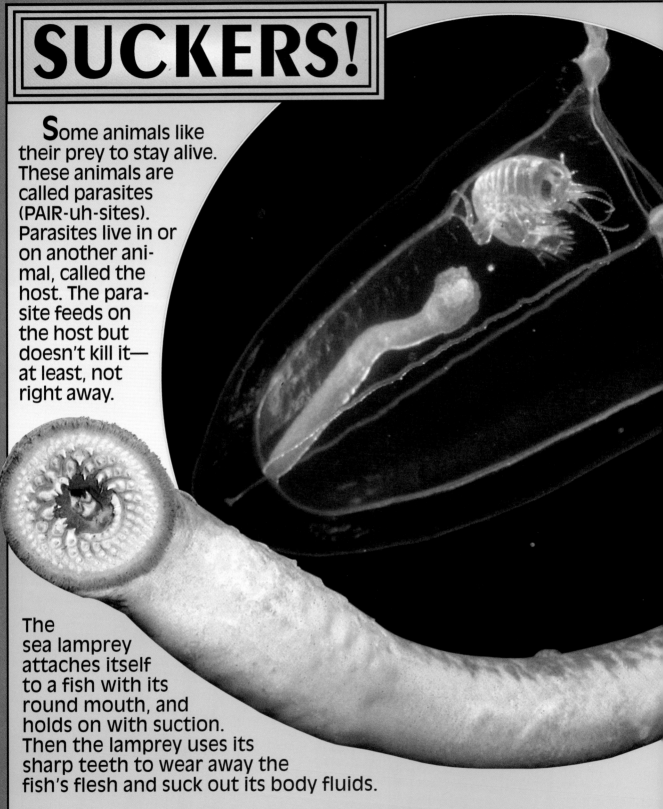

Some animals like their prey to stay alive. These animals are called parasites (PAIR-uh-sites). Parasites live in or on another animal, called the host. The parasite feeds on the host but doesn't kill it—at least, not right away.

The sea lamprey attaches itself to a fish with its round mouth, and holds on with suction. Then the lamprey uses its sharp teeth to wear away the fish's flesh and suck out its body fluids.

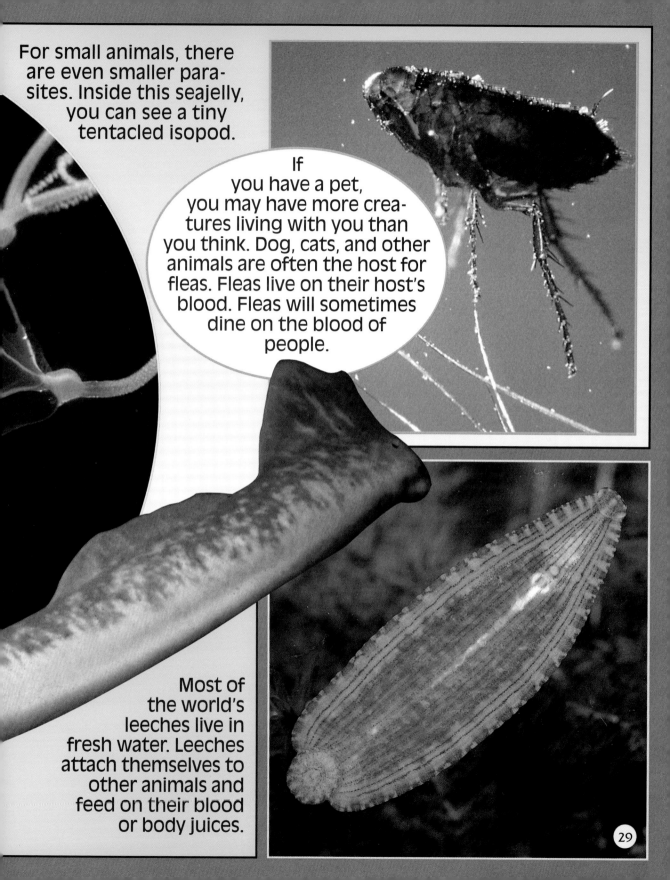

For small animals, there are even smaller parasites. Inside this seajelly, you can see a tiny tentacled isopod.

If you have a pet, you may have more creatures living with you than you think. Dog, cats, and other animals are often the host for fleas. Fleas live on their host's blood. Fleas will sometimes dine on the blood of people.

Most of the world's leeches live in fresh water. Leeches attach themselves to other animals and feed on their blood or body juices.

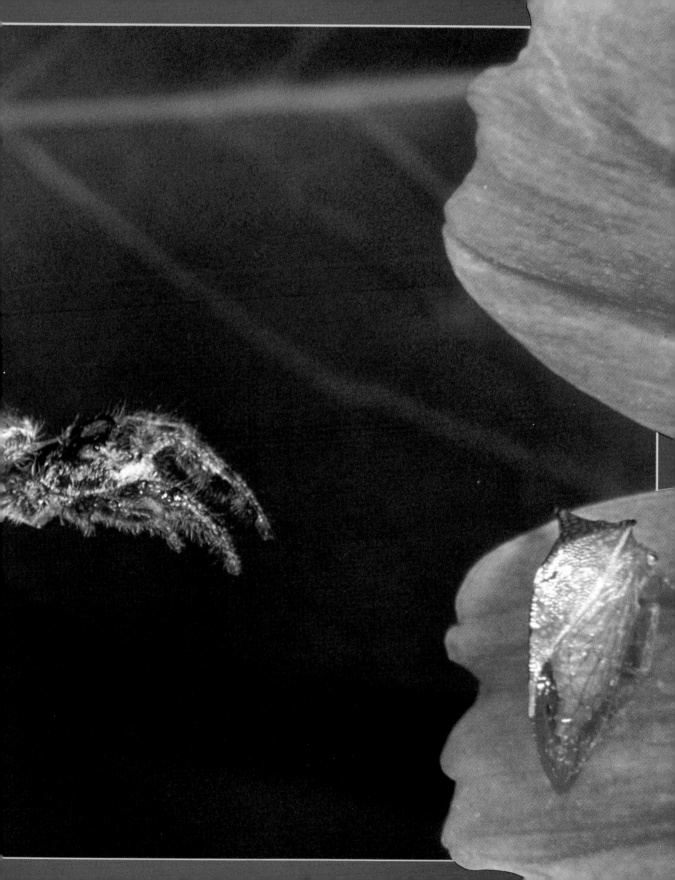